Creative Containers

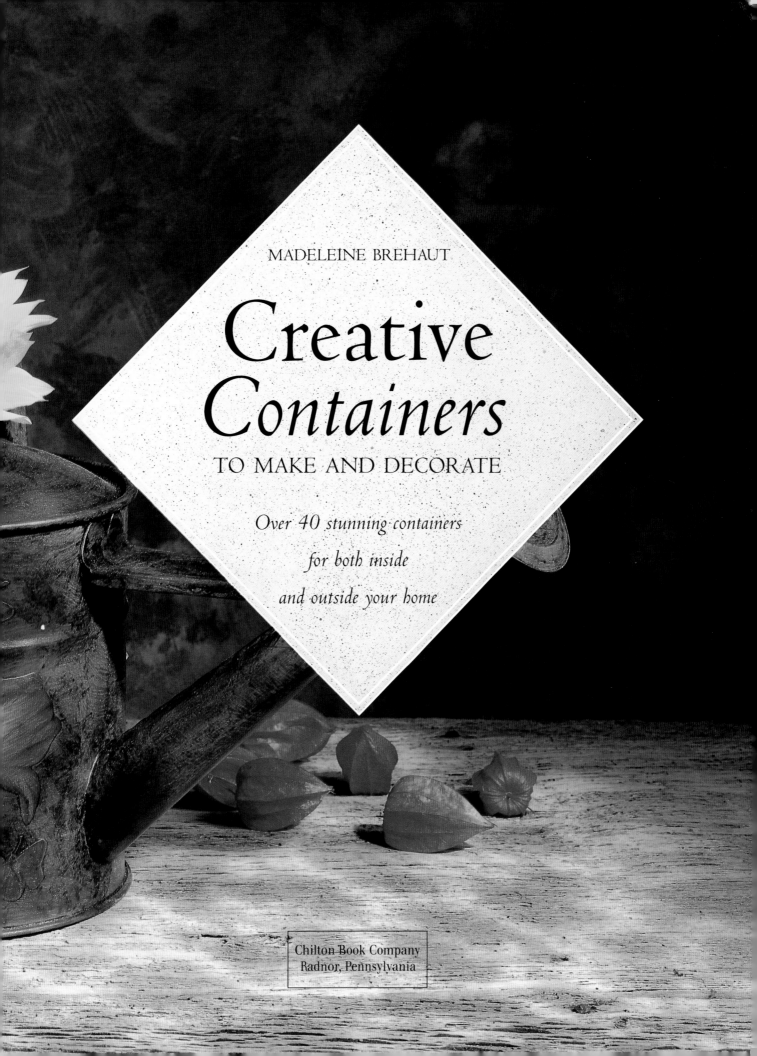

MADELEINE BREHAUT

Creative
Containers

TO MAKE AND DECORATE

Over 40 stunning containers

for both inside

and outside your home

Chilton Book Company
Radnor, Pennsylvania

A QUARTO BOOK

Copyright © 1997 Quarto Inc.

ISBN 0-8019-8941-8

This book was designed and produced by
Quarto Publishing plc
The Old Brewery
6 Blundell Street
London N7 9BH

Senior art editor Elizabeth Healey
Designer Tania Field
Editors Sarah Fergusson, Miranda Stoner,
Jo Fletcher-Watson, Jane Hurd-Cosgrave
Text editor Maggi McCormick
Managing editor Sally MacEachern
Photographer Martin Norris
Illustrators Dave Kemp, Jane Spencer
Backgrounds Steve Tse
Prop buyer Natalie Rule
Picture research manager Giulia Hetherington
Assistant art director Penny Cobb
Art director Moira Clinch
Editorial director Mark Dartford

Typeset in Great Britain by
Central Southern Typesetters, Eastbourne

Manufactured in Singapore by
Eray Scan Pte Ltd

Printed in Singapore by
Star Standard Industries (Pte) Ltd

Contents

Introduction

Since containers can be decorated in countless ways, they can add the finishing touch to any part of your home. They may be cheerful and bright, stylish and elegant, loud and outrageous, or quaint and rustic.

All areas of your home can benefit from a decorative container. The foyer can be given an elegant focal point with a decoupage vase filled with fresh flowers. A dull shelf can be brightened with a row of bright terracotta pots painted with chilies, etc., and containing cacti or other small plants. Even the bathroom could have a shell-covered bowl, filled with perfumed soaps or sandalwood.

There is a wide range of possibilities when you are creating containers for your home, so the forty projects in this book offer only a selection of suggestions. Once you have tried a few, you will begin to develop your own ideas, and may well adapt projects to suit your own home and lifestyle.

The book is divided into six chapters: terracotta; metal; china and glass; baskets; wood; and paper. There are many paint finishes within these chapters, but do not hesitate to incorporate one paint finish into a different project. The freestyle-painted herb pots would look delightful if given a craquelure finish. The decoupage folk-art box could be rag-rolled instead of being given a rubbed-paint finish. A stencil could be added to the plant container, or it could be marbled instead of dragged. The projects are merely a device to give you the confidence to start.

There are numerous styles within this book, and you will already have a preference. Whether it is contemporary, Victorian, country, or outrageously flamboyant, there should be a project to suit you.

The twig basket filled with terracotta pots and moss will give your kitchen windowsill a country-style look, and can be filled with small plants and dried arrangements of roses and lavender. The charming watering can looks like a treasured Victorian antique, but depending on your style, it could be filled with silk, dried, or fresh flowers. If you need an exotic, flamboyant display in the center of your table for an elegant dinner, the gold jeweled fruit stand will look superb when filled with tangerines, sugared fruits, or petit fours.

Some of these projects can be used to reinforce a theme, such as the patriotic stars and stripes box filled with dried gourds and corn, or the tartan-plaid box filled with thistles.

There is a distinction between containers that are suited to use indoors or out, but some of the projects are interchangeable. The paint finish usually governs the versatility of the project.

The Mediterranean mosaic pot filled with geraniums will bring the vibrant colors of that region onto your patio. Similarly, if the picket-fence box is protected with a waterproof varnish, it can be filled with pretty flowering plants to give an air of the countryside to any front door.

The two-tone candle box is really only suited to indoor use because of its contents, but if the paint finish is made waterproof, the box can be used outside.

The little stained-glass candle jars can decorate a patio or a dining room. However, the box covered with scratch-card paintwork should remain indoors, as should the stamped sunflower box or the hearts and flowers basket.

The whole aim of this book is to inspire you to begin creating and decorating your own containers. I hope it does, and good luck!!

Madeleine Brehaut

Basic *Techniques*

THE PROJECTS IN THIS BOOK ARE SIMPLE TO MAKE, AND DO NOT REQUIRE SPECIALIST SKILLS AND EQUIPMENT. IF YOU HAVE NOT ATTEMPTED ONE OF THE TECHNIQUES BEFORE, JUST PRACTICE ON A SMALL AREA OF PAPER, AN OLD PLATE, OR A PIECE OF WOOD (DEPENDING ON THE SURFACE NEEDED) UNTIL YOU ARE CONFIDENT TO START YOUR PROJECT.

CRACKLE GLAZE

This technique gives the effect of an aged and cracked paint finish.

1 Paint your project in a base color and let it dry. Apply the crackle glaze (there are many brands available), and let it dry.

2 Apply the top layer of paint. This can be either a similar or a contrasting color. As this layer of paint dries, it will crack, revealing the base color below. Varnish the surface once it has dried.

CRAQUELURE

The craquelure (delicately cracked) finish is achieved using two varnishes, which are sold in a pack. They dry at different speeds, causing very fine cracks to appear, which are barely visible until they are filled with colorant. The skill is in judging exactly when to apply the second varnish.

1 Apply a smooth and even coat of Varnish One, and let it dry until slightly tacky. Apply an even coat of Varnish Two, and let it dry completely. You can accentuate the cracks by heating gently with a hairdryer once the second coat has been applied to the surface.

2 The colorless cracks should now have formed. With a lint-free cloth, rub a little raw-umber paint over the whole surface. Once the cracks are filled, any excess pigment can be rubbed off. Seal with varnish.

STENCILING

Stencils can be bought already cut, or if you have a special design, it can be photocopied and then cut out of stencil paper.

1 Secure the stencil with a little masking tape.

2 Put a little acrylic paint onto a stiff artist's brush or a stencil brush. Dab off any excess onto a cloth before starting, and stipple it onto the surface with a quick, dabbing action. Apply each color required in turn, let them dry slightly, then remove the stencil.

MARBLING

This decorative paint technique is easier than it looks, and can add a touch of style and elegance to the plainest of containers.

1 Sponge the surface using a combination of your chosen background colors.

2 With a feather or fine brush, apply the veining color with a squiggle action.

3 Soften the veins with a wide soft brush. Let the paint dry and varnish for protection.

STAMPING

Patterned-rubber stamps are available from craft stores. Once the stamp is coated with ink, it is pressed onto a flat surface.

1 Load the stamp block with different-colored inks in turn, trying to accentuate the pattern.

2 Press the stamp block firmly, ink side down, onto the surface. Lift away, and repeat the process over your chosen surface.

SPONGING

This paint technique achieves a soft, muted finish, using a combination of complementary colors over a base color.

1 Dampen a natural sponge and load it with paint. Dab it a few times on a piece of paper or the edge of your palette to remove any excess. Start covering the surface in a random pattern, using a short, sharp, dabbing action and allowing some of the base color to show through.

2 Apply the second color in the same way, allowing the base color and the first sponged color to show through. Repeat the process with as many colors as desired.

Tools
and Equipment

PAINT BRUSHES
A selection of inexpensive artist's acrylic brushes in varying widths.

VARNISH BRUSH A sash brush 1–2in (25–50mm) wide.

SPONGE SPATULA A flat sponge on a stick handle found in hardware stores. It reduces the appearance of brush strokes to give a smoother finish.

ACRYLIC PAINT There are many brands of water-soluble acrylic paint. It is not necessary to have top-quality artist's paint; buy a less expensive kind.

CERAMIC PAINT A solvent-based paint used on china or glass. Try to find a type that does not need to be fired after use.

GLASS PAINT A solvent-based paint suitable for painting on glass.

LEADING Used to add a leaded appearance to designs when painting on glass. Available in gold, silver, and black.

ACRYLIC PRIMER Used as a base on metal or wood.

LATEX/ EMULSION PAINT A water-based paint used as a base coat.

VARNISH A protective sealant. Clear wood varnish can be colored with wood stains to add an aged appearance.

AEROSOL ARTIST'S VARNISH Matte or semi-gloss finishes are available. It is used on difficult areas for a smooth, even finish.

SCALPEL A craft knife used to cut paper and thin woods.

SCISSORS Paper-cutting, general quality.

CRAFT GLUE For sticking paper, wood, etc.

GLUE GUN Electric tool that melts special glue sticks that can be bought to suit different surfaces.

BALSA WOOD Fine, lightweight wood that cuts easily. Obtained from craft stores.

WAXED STENCIL PAPER Used to cut your own stencils.

PAPER AND CARDBOARD Selection of varying colors, weights, and styles.

OLD CERAMIC PLATE Can be used as a palette.

SANDPAPER Medium and fine grits.

SPONGES Natural or synthetic. A natural sponge will give a softer pattern.

LINT-FREE CLOTH Used for rag-rolling.

HAMMER Small decorator's type for general use.

RULER

DRAGGING BRUSH Available from art and craft stores, this brush has very long bristles that are more widely spaced than regular brushes.

FIXATIVE AND GROUT Used to attach ceramic pieces to terracotta. Once the fixative is dry, the spaces are filled with grout.

TERRACOTTA POTS Use frost-resistant types if they are intended to go outside.

BASKETS Available in all styles and materials, including wicker, cane, raffia, and reed.

FLORIST'S WIRE Thin wire available in different gauges.

RAFFIA Dried paper-like fiber available from craft stores and florists.

MOSS Different sorts are available from florists.

TERRACOTTA
HAS BEEN USED
THROUGHOUT HISTORY,
AND MUST BE ONE OF THE OLDEST
SUBSTANCES USED FOR CONTAINERS. THE
ANCIENT ROMANS USED TERRACOTTA URNS TO
CARRY WATER. ORIGINALLY, THEY WERE HAND THROWN,
BUT TODAY MOST ARE MACHINE MANUFACTURED. THEY HAVE A

Terracotta

SHINIER FINISH, AND SINCE THEY ARE MORE RESILIENT TO FROST, THEY CAN REMAIN OUTSIDE
FOR THE WHOLE YEAR. THE DESIGNS AVAILABLE ARE ENDLESS — SMALL, PLAIN, LARGE, OR
HIGHLY DETAILED. ● WHEN SEARCHING FOR POTS TO DECORATE, GO TO
GARDEN CENTERS AND THE MORE-USUAL OUTLETS, BUT ALSO TRY CRAFT
FAIRS FOR UNUSUAL DESIGNS, AND SECONDHAND STORES MAY HAVE
OLD, HAND-THROWN POTS THAT PEOPLE HAVE DISCARDED.
● WITH ALL THE PAINT FINISHES IN THIS CHAPTER,
IT IS IMPORTANT TO GIVE YOUR PROJECT A
COUPLE OF GOOD COATS OF
WATERPROOF VARNISH FOR
PROTECTION.

Freestyle-Painted Herb Pots

YOU WILL NEED
3 terracotta pots
White latex paint
Acrylic paint in magenta, cadmium
yellow, and cobalt blue
Opaque oxide of chromium green
Flat, spray-on varnish
Paintbrushes
Paper

THESE DELICATELY PAINTED PASTEL POTS WITH

FREESTYLE MOTIFS ARE MUCH SIMPLER THAN THEY FIRST APPEAR.

ONCE THEY ARE FILLED WITH WONDERFULLY SCENTED HERBS SUCH AS

BASIL, THEY WILL ADD A REAL

COUNTRY STYLE TO YOUR KITCHEN

WINDOWSILL.

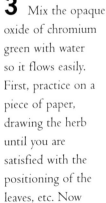

1 Cover the terracotta pots with white latex paint. If the brushstrokes are noticeable, it will add to the rustic effect. Let them dry.

2 Mix the magenta acrylic paint with the white latex until the desired pastel tone is achieved. Apply with strong brushstrokes, allowing a little of the white to show through. Repeat the process on the other two pots — one will be a pastel version of the cobalt blue; the other a pastel, cadmium yellow. Then let dry.

4 To paint the borders, apply a very small amount of paint on a wide brush, and lightly brush the top edge, adding just enough color to define the rim. For protection, the pots should be sealed with spray-on, flat, picture varnish.

3 Mix the opaque oxide of chromium green with water so it flows easily. First, practice on a piece of paper, drawing the herb until you are satisfied with the positioning of the leaves, etc. Now paint an herb on the face of each pot.

Stenciled *Flowerpot*

A PRETTY, HAND-CUT STENCIL OF IVY AND BUTTERFLIES ADORNS THIS SIMPLE TERRACOTTA POT. KEEPING THE TERRACOTTA POT ITS ORIGINAL COLOR ADDS TO ITS SIMPLICITY, BUT A COLOR WASH COULD BE USED FOR A BRIGHTER VERSION.

YOU WILL NEED
Terracotta flowerpot
Stencil film
Scalpel
Acrylic paints in opaque oxide of chromium, burnt umber, cadmium yellow, and venetian red
Paintbrush
Polyurethane varnish

18

1 Draw the butterfly and ivy, using the templates on page 122, onto the stencil film, then cut them out with a scalpel.

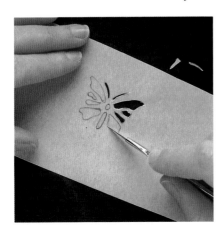

2 Mix the opaque oxide of chromium with a tiny amount of water. Place the ivy border stencil on the flowerpot and stencil the leaves. Then stencil the stems with the burnt umber. Continue until you have gone all the way around the pot, making sure that you butt the last section tightly up to where you started.

3 Now stencil the butterfly around the center of the pot, using the cadmium yellow and venetian red for the body. When paint is dry, add two coats of polyurethane varnish for protection.

Chili *Pot*

YOU WILL NEED
Terracotta pot
Selection of small paintbrushes
White latex paint
Acrylic paint in yellow, brown, green,
turquoise, and dark blue
Crackle varnish
Soft cloth
Beeswax polish or raw-umber paint
Flat spray-on varnish

THE RICH COLORS OF THIS TERRACOTTA POT ARE EVOCATIVE OF THE WARMTH OF THE MEDITERRANEAN. THE CHILIES DECORATE THE SIDES, BUT YOU COULD CHOOSE A FAVORITE FLOWER OR A STRIKING MOTIF OF YOUR OWN DESIGN IF YOU PREFER. THE SIDES OF THE POT HAVE BEEN CRACKLE-GLAZED TO GIVE ADDED CHARACTER.

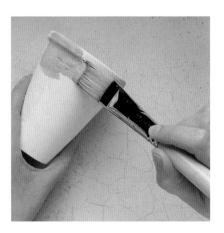

1 Cover the terracotta pot with a base coat of white latex paint. Make sure the outside of the pot is completely covered. Apply a top coat of yellow acrylic.

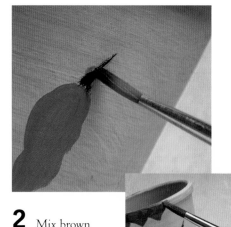

2 Mix brown and yellow acrylic paint to create the color for the main body of the chili motif. Add a stalk in green acrylic paint. (Practice first on paper if you wish.) Mix turquoise acrylic paint with a hint of blue, and use it to add a decorative design to the top and bottom of the pot. Let the borders dry fully.

3 Apply crackle varnish to the outside of the pot, following the manufacturer's directions.

4 With a soft cloth, rub in beeswax or raw-umber paint. For added protection, the pot can be sealed with flat spray-on varnish.

Terracotta *Pots*

BRIGHT PASTEL COLORS ARE ALWAYS POPULAR, AND THIS TUMBLING ARRANGEMENT OF TERRACOTTA POTS AND CANDLES WILL ADD A SENSE OF STYLE TO A BARBECUE OR OUTDOOR SUPPER. THE POTS ARE FILLED WITH SAND TO MAKE THEM STABLE AND SAFE.

YOU WILL NEED
Selection of terracotta pots
White latex paint
Acrylic paints in aqua, turquoise, cadmium yellow, cadmium orange, crimson, florescent green, and violet
Paintbrush
Candles of different sizes
Sand
Natural sponge

1 Paint the outside of all the terracotta pots with white paint and set them aside to dry fully.

2 Paint the large pot and one of the tiny pots with aqua and let them dry.

3 With the fluorescent green, paint random brush-strokes over the aqua. Now repeat with turquoise, allowing some of the aqua to show through, so that the finished effect is a random mixture of the three greens. Repeat the process on the tiny aqua flowerpot.

4 Paint two tiny terracotta flowerpots with violet acrylic paint and let them dry. Mix the crimson and a little white latex together to make a strong Mediterranean pink. Paint this pink onto one of the medium-sized pots, then add a few brush-strokes of pink to the tiny violet pots to soften the strength of the color. Let all the pots dry.

5 With the aqua, paint some random dots on the pink flowerpot. Then press softly on the surface with the sponge to give a mottled effect.

6 Paint one of the medium pots cadmium yellow and let it dry. Paint the last medium and tiny pots, cadmium orange, and add a few brushstrokes of cadmium yellow to give a two-tone effect. Once the yellow pot is dry, paint abstract V patterns in orange all over it.

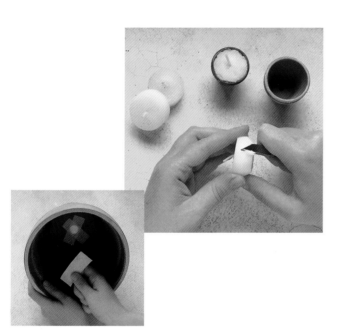

7 Trim the candles to fit the tiny flowerpots. Cover holes in bottom of pots with tape. Then put a large candle into the center of each medium pot and secure them with sand.

8 Fill the large pot with sand and arrange all the other pots by pressing them into the sand in a pyramid structure. Add a little more sand around them to make them secure.

Shell Pot

THE CASUAL ARRAY OF SHELLS AND STARFISH ON THIS COOL, VIOLET POT GIVES A HINT OF BALMY SEASHORE DAYS. FILL THE POT WITH A SCENTED CANDLE, AND PLACE IT IN THE BATHROOM FOR A GOOD, RELAXING, CANDLELIT SOAK!

YOU WILL NEED
Terracotta pot and saucer
Acrylic paint in light violet
Shells
Glue sticks and glue gun
Starfish
Sea lavender
Rye-grass string
Paintbrush

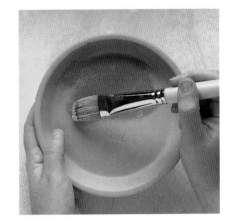

1 Paint the terracotta pot and saucer with a wash of light violet acrylic paint and let them dry.

2 Start gluing the rye-grass string to the painted pot, leaving the ends slightly free. Glue a starfish to the rye-grass string, and build up the design by adding shells in groups along the string.

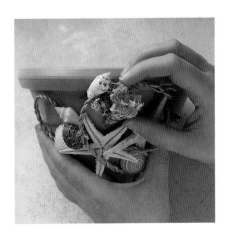

3 Continue adding starfish and shells until you have have gone around the pot. Finally, add small clusters of sea lavender. Fill with sand to support a candle, and place a few shells on top of the sand to decorate.

Verdigris *Terracotta Pot*

VERDIGRIS HAS BECOME

A VERY POPULAR PAINT

EFFECT RECENTLY, AND MOST KITS ARE VERY EASY TO USE.

AS MUCH OR AS LITTLE AS YOU WISH CAN BE USED,

DEPENDING ON THE COLOR DESIRED. HERE, HEAVY

RUNS OF LIGHT VERDIGRIS HAVE

BEEN ADDED TO THE POTS FOR

A MORE AGED APPEARANCE.

YOU WILL NEED
Terracotta pot
Acrylic paint in copper, turquoise green, aqua green, and white
Paintbrush
Natural sponge
Polyurethane varnish
Plate

1 Paint a base coat of copper onto the pot and repeat until the surface is completely opaque, allowing it to dry between coats.

29

2 Moisten a sea sponge with water and pour some of the turquoise green paint onto a plate. Fill the sponge with a little of the glaze. (Dab the sponge onto a piece of plain paper to check for pattern.) Start sponging the glaze onto the pot, making sure some of the copper still shows through. Allow some of the turquoise green glaze to run down the side of the pot.

3 Mix some aqua green and white paint with water and pour some of the glaze onto another plate area, then apply as before but covering less of the pot surface. Continue to let the glaze run down the pot to add to the aged appearance. Let it dry completely for 24 hours. Then protect it with polyurethane varnish.

Mosaic *Pot*

THE WILD, CLASHING

COLORS OF THESE

MOSAIC POTS ARE AS VIBRANT AS THE

TRADITIONALLY PAINTED FISHING BOATS THAT

DECORATE MEDITERRANEAN HARBORS. FILL THEM

WITH GERANIUMS TO ADD AN AIR OF SUMMER

TO YOUR HOME.

YOU WILL NEED
Terracotta pot
Assortment of different-colored
pottery shards
Tile adhesive
Grout
Lint-free cloth

1 Apply a thick,
even coat of adhesive to the side of the pot,
then comb some grooves through the
adhesive to allow for a better contact with
the shards of pottery. Work in a small area
at a time so you know that you can apply
the mosaic before the adhesive dries.

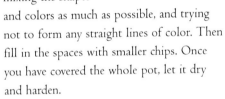

2 Start by covering the pot with larger pieces of broken pottery, mixing the shapes and colors as much as possible, and trying not to form any straight lines of color. Then fill in the spaces with smaller chips. Once you have covered the whole pot, let it dry and harden.

3 Grout the sides of the shards and push the grout into the spaces. Let the grout dry slightly, then wipe off the excess with a dry cloth. Once the grout has hardened, clean the pot with an abrasive cloth or very fine steel wool.

A DECORATED METAL CONTAINER MAKES AN UNUSUAL AND COLORFUL ADDITION TO A DULL CORNER, INSIDE OR OUT. ● FLORISTS AND HARDWARE STORES ARE USUALLY A GOOD SOURCE OF GALVANIZED-METAL CONTAINERS, AND GARAGE SALES AND SECONDHAND

Metal

STORES CAN BE SCOURED FOR ITEMS WAITING TO BE TRANSFORMED INTO TREASURES FOR YOUR HOME. OTHER METAL SHAPES INCLUDE OLD MILK CHURNS AND ENAMEL WATER PITCHERS. AS METAL IS DURABLE AND HAS A SMOOTH SURFACE, IT IS AN EASY-TO-DECORATE, LONG-LASTING, AND PRACTICAL CONTAINER.

Decoupage Watering Can

YOU WILL NEED
Metal watering can
White primer paint
Acrylic paint in leaf green and
Hooker's green
Gift wrap
Small sharp scissors
White craft glue
Clear satin varnish
Wood stain
Paintbrushes

THIS CHARMING WATERING CAN LOOKS AS THOUGH IT IS A TREASURED ANTIQUE. IN FACT, THE OUTSIDE OF THE CAN HAS BEEN DECORATED WITH FLOWERS AND BUTTERFLIES TAKEN FROM A SHEET OF GIFT WRAP. THE BACKGROUND OF THE CONTAINER HAS BEEN AGED WITH WOODSTAIN TO DARKEN IT.

1 Use white primer paint to cover the outside of the watering can.

2 Mix the leaf green and Hooker's green acrylic paints and apply it over the white primer paint on the watering can. Next, carefully cut out decoupage pieces from the gift wrap. Work out your arrangement before gluing them to the watering can.

3 Glue the two main pieces of the decoupage. We started with the large rose, then overlapped the smaller bloom around the can. Finish the decoupage with smaller details.

4 Use clear, satin varnish mixed with wood stain to darken the color and protect the finished watering can.

Punched Tin

THIS TRADITIONAL SHAKER-STYLE CRAFT HAS BEEN ADAPTED TO MAKE AN ATTRACTIVE PLANT HOLDER OR CANDY TIN, AND WITH THE AID OF VINEGAR, IT CAN BE AGED IN JUST 24 HOURS. ALL YOU NEED IS A CAKE PAN (AN OLD ONE WILL BE EVEN BETTER THAN A NEW ONE), A TIN PUNCH OR SHARP NAIL, AND SOME VINEGAR!

YOU WILL NEED
Cake pan
Tin punch (or sharp nail)
Hammer
Paper towels
Vinegar
Paper and pencil
Scrap of wood

1 First, draw or trace a design on paper. Once you are happy with the arrangements and dimensions, draw the design on the side of the cake pan with a soft pencil or sharp nail.

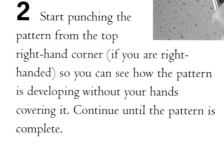

2 Start punching the pattern from the top right-hand corner (if you are right-handed) so you can see how the pattern is developing without your hands covering it. Continue until the pattern is complete.

3 Soak paper towels in vinegar and lay them on the side of the cake pan for half an hour. Remove them for a few minutes to allow the air to attack the metal, then replace the vinegar-soaked towels. Repeat until you have the level of aging that you desire.

Stenciled Bucket

ADD A TOUCH OF COLOR AND FUN TO YOUR

PATIO OR HOUSE WITH THIS METAL BUCKET, WHICH

HAS BEEN STENCILED WITH DAISIES, IRISES, AND

TULIPS TO GIVE IT A FOLK-ART STYLE.

YOU WILL NEED
Green metal bucket
Acrylic paints in fluorescent blue,
brilliant yellow, process magenta,
cadmium scarlet, opaque oxide of
chromium, Winsor purple, cerulean
blue, and white
Assorted stencils (page 123)
Paintbrush
Plate or palette

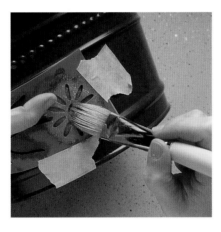

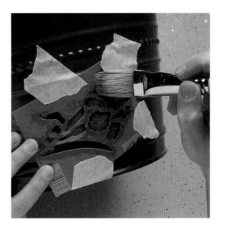

1 On a plate or palette, mix a little fluorescent blue and brilliant yellow (to make green) and place a small amount of all the other colors at intervals around the plate. This is now your palette. Start by stenciling the daisies onto the center of the bucket. The colors used are process magenta, scarlet, and the white. The leaves are painted in the two greens with highlights of yellow.

2 Place the iris stencil next to the daisy and stencil it with the blue and purple, then add the greens for the leaves. Continue stenciling the iris and daisy alternately.

3 Around the top edge of the bucket, stencil the tulip border using the cadmium scarlet and yellow alternately. Highlight the ridged dots on the bucket in blue and the line bands in yellow. Finally, paint a yellow line along the center of the handle.

Print *Vase*

YOU WILL NEED
Galvanized bucket
Acrylic primer
Yellow-gold acrylic paint
Sponge spatula
Craft glue
Paintbrush
Black-and-white flower prints
Spray varnish

THIS ELEGANT PRINT VASE IS REALLY A GALVANIZED BUCKET! IT HAS BEEN PAINTED WITH YELLOW-GOLD ACRYLIC PAINT AND DECORATED WITH PHOTOCOPIED DECOUPAGE GARLANDS, BORDERS, AND FLOWERS.

1 Paint the bucket with two coats of acrylic primer. Let it dry. Apply two coats of yellow-gold acrylic paint using the sponge spatula to avoid paint ridges. Let the first coat dry fully before applying the second.

2 Photocopy the flower prints and cut them out.

3 Add water to the glue to make it easy to apply with a brush. Start with the top border and butt up each strip of pattern close to the previous piece. Apply the bottom border in the same way. Glue the garlands of flowers to the top of the bucket just below the top border. Finally, add one basket of flowers centrally beneath each garland swag. Let it all dry.

4 Apply a coat of diluted glue to the bucket. Let it dry, and then spray on several coats of varnish.

YOU WILL NEED
Wire coat hanger
Chicken wire
Wire cutters
Large flowerpot containing ivy plant

1 Unwind a coat hanger at the hook, then cut the long straight side in the middle, making two pieces of wire. Take the wire piece with the hook on it, and fold the long edges closer together to form a pair of legs; then bend the hook parallel to the ground to make a base.

Wire *Peacock*

THIS IDEA IS REMARKABLY SIMPLE, AND IS EXTREMELY REWARDING ONCE YOUR PLANT HAS STARTED TO FLOURISH. IT CAN BE ADAPTED TO BECOME A CHICKEN, A DUCK, OR EVEN A PIG, AND BECAUSE OF THE TACTILE NATURE OF THE WIRE, YOU CAN EASILY RE-SHAPE IT IF YOU MAKE A MISTAKE.

2 Take the other half of the coat hanger, and bend the twisted part of the wire that was originally wrapped around the neck of the hook into a concertina three times to make the coronet. The twist at the end makes the beak, while the other straight end of the wire forms the neck support.

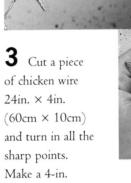

3 Cut a piece of chicken wire 24in. × 4in. (60cm × 10cm) and turn in all the sharp points. Make a 4-in. cut in the center of one end with wire cutters. Fold in the sharp edges, then wrap the two strips of chicken wire that you have cut around the beak and coronet, molding them into a head.

4 Cut another piece of chicken wire 12in. × 24in. (30cm × 60cm) and turn in all sharp edges. Turn in two corners at one end, placing the neck between the folded corners, then bend them up and over the neck to make the body. Curl the other ends of the wire sheet together to form the tail. Now curl the end of the neck into a high arch to make the top of the bird's back.

5 Thread the legs through the bird's chest and place them in the flowerpot. Push ivy shoots inside the bird; as they grow, trim back to the wire.

This chapter holds a combination of mediums — china and glass. ● Historically, ceramics have always been more expensive than other household materials. In fact, in the eighteenth century, many homes would have had no china items apart from a white china pitcher. Nowadays, china is relatively inexpensive and readily

China and Glass

available. For these projects you can use old, new, chipped, faded, or discolored china. Every home — even the neatest — has a collection of unwanted and never-used china. ● For me, one of the delights of craft work is sourcing the bare materials to work with, and glass is always a pleasure, as it often involves jams, mustards, preserves, and other delicious things. Glass is produced in so many colors, designs, and styles that the possibilities are endless.

Ceramic Painted Pitcher

YOU WILL NEED
Pitcher
Ceramic paints in garnet red,
kingfisher blue, vermilion, and yellow
Paintbrush
Polyurethane varnish

BRIGHT, GARISH, NAIVE FLOWERS AND WACKY STRIPES AND SPOTS HAVE TRANSFORMED THIS DULL OLD PITCHER INTO A STYLISH DESIGNER VASE. THE CERAMIC PAINTS DO NOT HAVE TO BE FIRED, BUT A COUPLE OF COATS OF VARNISH WILL PROTECT THE PAINT.

48

1 Paint the flowers on the pitcher in garnet-red paint, using a regular brick work pattern.

2 Paint a dot in the center of the flowers in kingfisher blue, then let the paint dry.

3 Paint vermilion lines vertically around the pitcher at regular intervals. Do not worry if they are a little wobbly; this adds to the design. Finally, paint yellow dots down the vertical line spaces. Let it dry completely for at least 24 hours, and then varnish with polyurethane varnish.

Marble *Ceramic Bowl*

YOU WILL NEED
Plain ceramic bowl
Sea sponge
Poppy-red glaze with pale-pink
base coat
Fine paintbrush
Large, soft paintbrush
Polyurethane varnish
Plate

MARBLING LOOKS VERY DIFFICULT, BUT WITH JUST A LITTLE PRACTIC

A VERY PROFESSIONAL FINISH CAN BE ACHIEVED.

A REALISTIC ROSE COLOR HAS BEEN USED TO COVER TH

CERAMIC BOWL, BUT YOU CAN ALWAYS TRY SOMETHING

MORE FANCIFUL, SUCH AS ORANGES OR PURPLES.

1 Clean and dry the bowl thoroughly so that there are no grease marks. Dampen the sea sponge. Pour a little of the pale-pink base coat onto a plate, dip the sponge in it, and gently sponge the base and the inside of the bowl – try to cover nearly the whole surface.

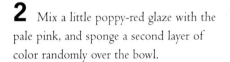

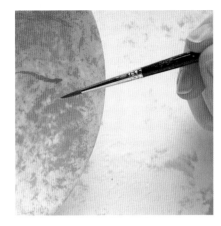

3 With a very fine paintbrush, draw wiggly lines in the poppy glaze. Soften them with the larger brush so they become fairly fuzzy. Keep the lines random, but remember that all the veins in marble run basically the same way; they do not form a crisscross pattern. Veins can also be painted on the base of the bowl for a more realistic effect. Finally, when the glaze is completely dry (24 hours), varnish with several coats of polyurethane varnish.

2 Mix a little poppy-red glaze with the pale pink, and sponge a second layer of color randomly over the bowl.

Sponged
Bowl Ceramic

YOU WILL NEED
Ceramic bowl
Ecru base coat
Sea sponge
Pumpkin translucent glaze
Sunshine translucent glaze
Polyurethane varnish
Plate

SPONGING IS ONE OF THE EASIEST PAINT FINISHES TO ACHIEVE. HERE, COORDINATED COLORS HAVE BEEN USED TO GIVE A SUBTLE, YET VIBRANT LOOK. SEVERAL COLORS COULD BE ADDED TO GIVE A MORE COMPLEX, MARBLED FINISH.

1 Shake the ecru base coat for at least one minute so it is completely mixed, then pour a little onto a plate. Dampen the sea sponge and cover the whole bowl with the paint, trying to leave as little white showing as possible. Let it dry.

2 Shake the pumpkin glaze and pour a little onto a plate. Randomly cover the bowl, allowing the ecru base coat to show through. Let it dry until it is tacky.

3 Finally, mix the sunshine glaze and pour some onto a plate, then apply it over the pumpkin, blending it in so the resulting color becomes muted and fuses together. Let it dry completely for 24 hours, then varnish with polyurethane varnish.

Red Oxide– and Gold-Glazed Ceramic Vase

YOU WILL NEED
Ceramic vase
Metallic-gold glaze and
red-oxide base coat
Paintbrush
Sea sponge
Aerosol polyurethane varnish

THIS LUXURIOUS VASE IS JUST PAINTED IN A BASE COAT OF DEEP-RED OXIDE AND SPONGED SEVERAL TIMES WITH A GOLD GLAZE FOR A GLORIOUSLY RICH DISPLAY.

1 Paint the vase with the red-oxide base coat. Let it dry, and repeat until no white can be seen through the paint.

2 Shake the gold metallic glaze thoroughly for at least one minute, then pour some onto a plate. Dampen the sea sponge. Begin sponging the gold randomly around the vase, making sure the red oxide still shows through. Let it dry, then repeat so that the gold glaze really shines over the base coat. Let it dry, and then spray on varnish.

Stained-Glass Jars

YOU WILL NEED
Selection of glass jars
Gold leading
Stained-glass paints in crimson,
orange, chartreuse
Paintbrush

THESE JEWEL-COLORED STAINED-
GLASS NIGHT LIGHTS ARE JUST SIMPLE JARS ONCE
FILLED WITH MUSTARDS AND PRESERVES. THEY WILL
DECORATE YOUR PATIO OR YARD, AND BE PERFECT FOR LONG, BALMY,
SUMMER EVENINGS. FILLED WITH SMALL CANDLES OR WITH COLORED
WATER AND FLOATING CANDLES, THEY ARE IDEAL FOR OUTDOOR
SUMMER ENTERTAINING. USE STRING OR WIRE TO
HANG THEM UP IF PREFERRED.

1 With the gold leading, draw the pattern of scallops and flowers onto the glass jar and let it dry.

2 Paint the scallops at the top and bottom of the jar in crimson. Ensure the glass is covered with paint.

3 Fill in the flowers with orange paint, and complete the jar by painting the background chartreuse. This is a simple pattern for you to start with, but as you can see, the designs are endless. Silver and black leading can also be obtained, and there is a wide variety of paint colors available.

Painted-*Glass Tumblers*

TURN OLD TUMBLERS INTO FUN-COLORED, MODERN VASES USING GLASS PAINTS. THEY CAN BE DECORATED AS THESE ARE, WITH FLOWERS AND GEOMETRIC PATTERNS, OR YOU COULD COVER THE GLASS COMPLETELY WITH AN ELABORATE DESIGN, OR EVEN THEME THEM TO A SPECIFIC EVENT.

YOU WILL NEED
Glass tumblers
Stained-glass paint in orange, chartreuse, lemon yellow, and crimson
Paintbrush
Mineral spirits

1 Clean and dry the glass completely, making sure there are no greasy patches. Paint on the large, orange spots in a regular pattern. If any of the paint runs, clean the area with mineral spirits.

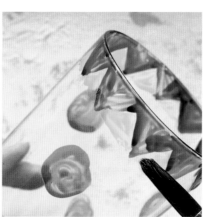

2 Paint chartreuse triangles at the top and bottom of the glass. Let it dry, then fill in the triangular spaces with lemon-yellow paint.

3 Finally, with single brushstrokes, paint crimson lines around the orange spots, completing the suns. Let it dry for 24 hours, until the paint is hard.

YOU WILL NEED
Gold base coat and copper
translucent glaze
Sea sponge
Glass-stemmed dish
Spray varnish
Plastic "gemstones"
Glue
Gold-colored glass leading
Plate

Gold Jeweled Fruit Stand

THIS DECADENT FRUIT STAND HAS BEEN SPONGED IN SEVERAL COLORS OF GOLD ACRYLIC PAINT AND THEN ADORNED WITH PLASTIC "GEMSTONES" OBTAINED FROM CRAFT AND HOBBY STORES. IT WILL LOOK SPLENDID ON A DINING-ROOM TABLE, ESPECIALLY DURING THE HOLIDAY SEASON, IF FILLED WITH TANGERINES.

1 Shake the base coat vigorously and pour a little into a plate. Dampen a sea sponge and apply the gold liberally all over the dish, including the inside. Let it dry, and then repeat the process with the copper glaze. Let it dry again, then spray varnish over the whole dish and set it aside for 24 hours.

2 Glue the gemstones around the bowl and base of the dish.

3 With gold leading, draw a line around the gemstones and decorate the top and bottom edge of the glass with a filigree pattern of swirls and dots, using the gold glass leading.

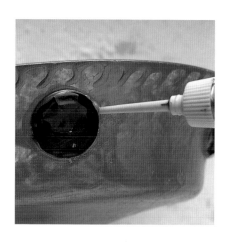

Tortoiseshell *Vase*

You will need
Glass vase
Gold base coat
Indian-brown glaze
Black glaze
Ordinary varnish
Paper plates
Paintbrushes
Toothbrush
Spray varnish

THIS IMPOSING TORTOISESHELL VASE WILL MAKE A DRAMATIC STATEMENT IN YOUR HOME. ALTHOUGH IT IS A LITTLE TIME-CONSUMING, IT IS DEFINITELY WORTH THE EFFORT.

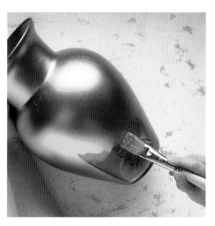

1 Make sure the vase is clean and dry and paint it all over with the gold base coat. Apply several coats until an opaque coverage is achieved. Let it dry between each coat.

2 Shake the glazes well and pour some Indian-brown glaze into a paper plate. In a different area of the plate, pour equal parts of the Indian-brown and black glaze and mix. In a third area, pour just the black glaze.

3 Brush a coat of varnish over the vase.

4 Begin placing small oval-shaped dabs of the brown-black paint mixture on the vase. Try to make a fan shape, as if each dab is part of a shell. Add some Indian-brown dabs next to or over the brown/black dabs.

5 Using a wide, soft brush, slightly drag the dabs so they blur into the wet varnish.

6 Continue building up the pattern, adding a few, plain-black dabs as well.

7 Put a little of the dark-brown mixture on an old toothbrush. Then, gently brush the bristles over the vase with downward strokes to create a tortoiseshell effect.

8 Varnish with spray varnish, and let dry for 24 hours.

BASKETS HAVE
BEEN USED AS
CONTAINERS FOR CENTURIES.
THE CHOICE OF STYLE, SIZE, AND
COLOR IS NOW SO VARIED THAT YOU
SHOULD HAVE NO PROBLEM IN FINDING THE
PERFECT SHAPE AND STYLE TO SUIT YOUR PROJECT. IF THE
BASKET HAS AN UNWANTED HANDLE, JUST CUT IT OFF. EVEN OLD

Baskets

AND WORN BASKETS HAVE CHARM AND BEAUTY IN THE RIGHT SETTING.   ●   YOU CAN FILL
THE BASKET WITH FLORIST'S FOAM TO SUPPORT YOUR DISPLAY, OR EVEN HIDE A WELL-
WASHED FOOD CAN IN YOUR BASKET TO HOLD FRESH-CUT FLOWERS.   ●   MOST
OF THE DECORATIONS IN THIS CHAPTER ARE NATURAL INGREDIENTS, SUCH
AS NUTS, PINE CONES, SHELLS, TWIGS, AND WHEAT, WHICH CAN BE
GATHERED DURING A PLEASANT WALK ALONG THE BEACH OR
IN THE COUNTRYSIDE.   ●   I HAVE USED A GLUE
GUN IN THIS CHAPTER, BUT IF YOU DO NOT
OWN ONE, DON'T GO OUT AND BUY
ONE. A GOOD-QUALITY
ADHESIVE WILL DO THE
JOB JUST AS
WELL.

Fir Cone *and* Nut Basket

YOU WILL NEED
Basket
Cinnamon sticks
Raffia
Small flowerpots
Synthetic batting
Selection of seeds and pods
Glue gun and clear-glue sticks
Fir cones
Selection of nuts

THE BEAUTY OF THIS BASKET IS THAT IT CAN BE DECORATED WITH SUCH A VARIETY OF ITEMS. YOU CAN USE THE SEEDS, PODS, AND NUTS THAT WE'VE USED HERE, OR TRY THINGS THAT CAN BE COLLECTED ON A WALK, OR EXTRA INGREDIENTS FROM YOUR KITCHEN CUPBOARD, SUCH AS DRIED FRUIT, CHILI PEPPERS, OR DRIED MUSHROOMS.

FOR A HOLIDAY FEEL, SPRAY SOME OF THE CONES WITH A HINT OF GOLD, BRONZE, OR SILVER.

1 Bundle the cinnamon sticks into groups of six and tie them with raffia.

2 Fill the small flowerpots with just enough batting to reach the pot rim. Place a selection of seeds and pods in a saucer, then coat the batting with a generous amount of glue and roll the top of the flowerpot over the seeds, until none of the batting shows (a little extra glue may be added to make sure the seeds really stick together).

3 Start decorating the basket with fir cones, then add a selection of nuts, layering them to add interest. Place a bundle of cinnamon sticks on the rim of the basket. Keep adding pieces in separate groups until you have covered the whole basket.

Twig Basket

THIS TWIG BASKET CAN BE MADE IN ANY SIZE TO MATCH POTS THAT YOU ALREADY HAVE OR WOULD LIKE TO DISPLAY IN SUCH A DECORATIVE WAY. THIS PROJECT COSTS VERY LITTLE, AND WOULD MAKE A CHARMING GIFT TO CHEER UP ANY WINDOWSILL.

YOU WILL NEED
Selection of pots
Selection of dry twigs
Strong garden scissors
Florist's wire
Moss
Glue

1 Choose the pots you would like to make a basket for and place them in a row, leaving a ½-in. (1cm) space between each one. Lay a dry stick along each long side of the row of pots and cut them to length, allowing an extra 1½in. (3cm) at both ends.

2 Do the same for the short sides and between each pot, laying these shorter sticks on top of the longer ones. Now wire the sticks together so they start to form a frame around the pots. Remove the pots and wire two or three longer sticks across the bottom of the frame to form a base. Build up the frame by laying sticks on top of each layer and wiring as you go until you reach the height of the pots.

3 Place the filled pots in the basket frame and decorate with moss to finish.

Hay Rope Basket

YOU WILL NEED
Hay
Raffia
Darning needle
Scissors

THE HAY ROPE USED IN THIS PROJECT MAKES A SWEET NEST TO SET OFF AN ARRANGEMENT OF SPRING-FLOWERING BULBS, OR AN EASTER GIFT FILLED WITH EGGS AND FEATHERS.

1 Take a good handful of hay and roll it into a sausage shape. Start by wrapping the raffia several times around the base and tying it firmly. Now twist the raffia firmly around the hay in a spiral (make sure that it stays evenly spaced, with about ½in. [1cm] between each raffia turn). Keep adding more hay until the required length of rope is made. For this basket, it was 8ft (3m).

2 Thread the darning needle with a double length of raffia. Tie the loose end firmly around the beginning of the hay rope. To start a spiral base, fold the end of the rope back onto itself and sew it securely (try to follow the spiral already formed by the bound raffia). Coil the rope around a little more and sew it securely onto the previous coil. Continue sewing the spiral until the desired base size is formed.

3 Now start to spiral the hay rope up to form the sides of the basket, still sewing in each twist. Continue until all the rope is sewn down. Fasten off the end by knotting the raffia once, and then sew back into the spiral several times. Trim any loose strands.

Wheat Basket

YOU WILL NEED

Acrylic paint in yellow ocher, deep cadmium red, and leaf green

Basket

Wheat

Barley

Raffia

Florist's decorations

Glue

Paintbrush

THE MUTED TONES OF EVER-POPULAR WHEAT, BARLEY, AND RAFFIA DECORATE THIS PAINTED BASKET AND ADD A RUSTIC, COUNTRY FEEL TO ANY HARVEST TABLE SETTING.

1 Mix equal parts of yellow ocher and leaf-green acrylic paint, and roughly paint the basket, allowing some of the willow to show through. Let it dry. Mix one part yellow ocher to two parts deep cadmium red, and then paint a second coat to give the basket a two-tone effect.

2 Tie the wheat and barley into tight bundles with the raffia. Make them different lengths so they form the shape of a fan as required to cover the sides of the basket. Loop an extra length of raffia and tie it to give the impression of a bow.

3 Glue the bundles of wheat and barley to one side of the basket, starting at the lower edge and finishing at the center with the shortest bundles. Now add the raffia bundles, and finish with the florist's decorations. Repeat on the other side.

Wicker Tray and Glass Holder.

YOU WILL NEED
Round, wicker tray
4 wicker glass holders
White latex paint
Acrylic paint in cobalt blue and aqua
Sponge
Craft knife

THESE SUNNY

CHECKERBOARD-PAINTED WICKER GLASS HOLDERS

WILL BRIGHTEN UP A PICNIC, AND THE TRAY

CAN DOUBLE AS A CHEESE PLATTER.

1 Paint the wicker tray and four glass holders with white latex and let them dry.

2 Mix the cobalt blue with a little white and paint the tray, allowing some of the white to show through.

3 Mix the aqua with a little white and then paint the tray again, leaving some of the blue and white to show through.

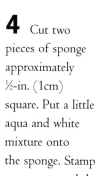

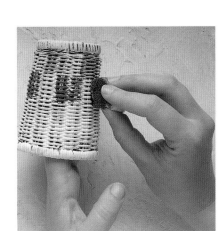

4 Cut two pieces of sponge approximately ½-in. (1cm) square. Put a little aqua and white mixture onto the sponge. Stamp squares around the base of the wicker at regular intervals, leaving a white square area between each aqua square.

5 Put the cobalt blue-and-white mixture on the second sponge and stamp above the aqua line, positioning the blue square above the white area. A checked pattern should start to form.

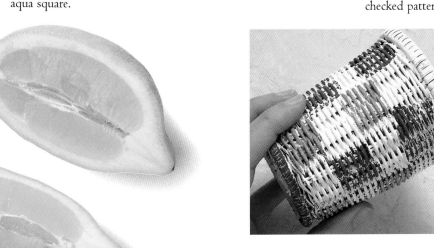

6 Repeat pattern once more with the aqua and then again with the blue. Four lines should cover the side of your wicker holder.

7 Paint a blue-and-white line around the very base of the glass holder, and then an aqua-and-white line on the rim of the holder. Repeat the whole process on the remaining three holders. Let them dry. When these are dry, place the glass holders inside the wicker holders.

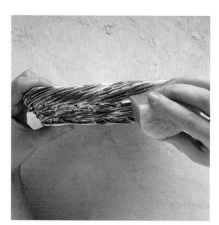

8 Sponge on varnish to both the tray and the four holders.

YOU WILL NEED
Twig basket
Balsa wood
Pencil
Scalpel
Acrylic paints in deep cadmium red,
renaissance gold, cadmium yellow,
leaf green, venetian red, and purple
Paintbrush
All-purpose glue
Rye grass

Hearts and Flowers Basket

HEARTS AND FLOWERS HAVE BEEN BRIGHTLY PAINTED AND USED TO DECORATE THIS TWIG BASKET IN A FOLK-ART STYLE. FOR AN ALTERNATIVE, SHELLS, STARFISH, BOATS, AND EVEN HOUSES COULD BE USED, DEPENDING ON THE CONTENTS OF THE BASKET AND ITS POSITION IN THE HOME.

1 Draw six large hearts, two small hearts, and six flowers on balsa wood, using the templates on page 124. Cut out all the shapes with a scalpel.

2 Paint the hearts deep cadmium red. Add a few brushstrokes of renaissance gold when the red is dry. Paint the large, star-shaped flowers cadmium yellow, adding brushstrokes of leaf green and a speckled center of venetian red. Finally, paint the other flowers purple with a cadmium-yellow center and leaf-green leaves.

3 Glue three large hearts along each long side of the basket, and arrange the two yellow star-flowers on top of a few rye-grass stems. One small heart goes on each end of the basket, with a purple flower and green leaves. The basket can be lined with dry moss or a pretty fabric, and then filled with fresh or dried flowers.

Country *Hay Nest*

YOU WILL NEED
Dry florist's foam block
Fabric
Several pins
Small- or medium-gauge florist's wire
Hay
Rye-grass rope

THIS CHARMING, RUSTIC HAY NEST SETS OFF DRIED FLOWERS AND GRASSES SUPERBLY. IT IS VERY QUICK AND EASY TO ACHIEVE, AND THE SECRET OF THE WHOLE DISPLAY IS ITS SIMPLICITY. IT CAN BE MADE IN SEVERAL SIZES FOR DIFFERENT EFFECTS. A VERY SMALL BASKET FILLED WITH EASTER EGGS WOULD BE SWEET, AND A LARGER ARRANGEMENT OF TALL WHEAT AND OTHER GRAINS WOULD LOOK IMPRESSIVE IN A FIREPLACE.

1 Cut a piece of florist's foam, and then cut the fabric large enough to cover the base and sides of the block. Fold it neatly around the block so there are no wrinkles and anchor it with pins.

2 Wrap the wire around the foam block several times to secure it. Add clumps of hay to the sides while wrapping the wire around the block. Continue until all sides are covered. Do not worry about how messy it looks. Fasten off the wire by pushing it under the wrapped wire and winding it over and over itself. Trim the hay at the base so it will sit level on a table.

3 Wrap the rye-grass rope around the block, covering the wire as much as possible. For a rustic effect, finish with several knots, and coil the rope back into the hay, then insert into the nest. Fill with dried grasses and flowers.

NOT EVERYONE
IS A MASTER
CARPENTER AS WELL AS A
CRAFTSPERSON — SO IN THIS BOOK WE
HAVE FOUND WOODEN "BLANKS" THAT ARE
AVAILABLE FROM GOOD CRAFT STORES. HOWEVER,
YOU MAY FIND SOME "BLANKS" THAT HAVE BEEN
DISCARDED — FOR EXAMPLE, AN OLD JEWELRY BOX, A WOODEN

Wood

FRUIT CRATE, OR A WINE CASE.   ●   ALL THE PROJECTS USING RAW WOOD WILL NEED TO BE
WELL SANDED AND PRIMED WITH ACRYLIC PRIMER BEFORE USE TO ENSURE A GOOD
BONDING OF PAINT, WITHOUT LOSING THE VIBRANCY OF ANY COLORS.   ●
REMEMBER THAT IF YOU PLAN TO PLANT INSIDE A WOODEN CONTAINER,
YOU WILL NEED TO FIT A WATERPROOF LINER INSIDE TO PREVENT
THE WOOD FROM ROTTING.

Picket-fence Box

YOU WILL NEED
Wooden crate
Balsa wood
Burnt-sienna acrylic paint
Crackle glaze (plaid)
Sponge spatula or brush
White latex paint
Flat, all-weather varnish
Wood glue
Craft knife

THIS PICKET-FENCE BOX LOOKS LIKE IT HAS BEEN WEATHERED OVER MANY YEARS, BUT IT IS ALL-NEW MATERIAL. THE GLAZE IS READILY AVAILABLE, AND WORKS INSTANTLY. FILL THE BOX WITH COLORFUL FLOWERING PLANTS TO COMPLEMENT THE COTTAGE STYLE OF THE PICKET FENCE.

1 Using the craft knife, cut out the number of slats from the balsa wood you need to cover the sides of your crate.

2 Paint the slats and the crate with the burnt sienna acrylic paint. Let them dry.

3 Glue the slats to the side of the crate at regular intervals. Let them dry. Use the sponge spatula or brush to apply the crackle glaze with vertical strokes.

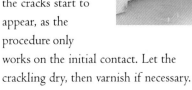

4 Apply white latex to the slats in the same direction as the crackle glaze. Try not to touch the paintwork once the cracks start to appear, as the procedure only works on the initial contact. Let the crackling dry, then varnish if necessary.

Stars *and* Stripes

YOU WILL NEED
Small wooden box
Balsa wood
Acrylic paint in red oxide and
ultramarine blue
All-purpose glue
Craft knife
Small paintbrushes

This patriotic box can add a touch of tradition to your home. It is very simple to make and can be extended to carry a theme to a dinner party table. It would also make an original gift-box idea.

1 Draw a five-pointed star on balsa wood and cut it out with the craft knife. Use it as a template to cut another four stars so you have five stars in all. Colorwash them with ultramarine blue acrylic paint, and let them dry completely.

2 With red-oxide acrylic paint, drag rough lines horizontally across the box, trying to leave a little wood showing through. Repeat on all four sides of the box, and let it dry.

3 Glue the stars to the front and sides of the box, using a random pattern.

Folk -art Box

YOU WILL NEED
Wooden box
White latex paint
Acrylic paint in cobalt blue
Cranberry glaze
Lint-free cloth
Folk-art prints
Glue
Varnish
Paintbrush

THE TRADITIONAL FOLK-ART STYLE OF THIS PRETTY BOX HAS BEEN ACHIEVED USING A PRINTED DECOUPAGE KIT. SOME OF THE CRANBERRY GLAZE HAS BEEN RUBBED AWAY TO AGE THE APPEARANCE OF THE BOX.

1 Paint the box and lid with white latex and let it dry. Then paint the box and lid with cobalt blue acrylic paint and let it dry.

2 Paint the outside of the box and lid with cranberry glaze. Rub a little away with the lint-free cloth.

4 Apply several coats of spray varnish to the box to finish it.

3 Cut out the folk-art prints and glue them to the sides and lid of the box, pressing down firmly.

Candle *Box*

YOU WILL NEED
Wooden candle box
Acrylic paints in crimson, aqua, and
cobalt blue, plus small amounts of
yellow ocher, cadmium red, and
venetian red
Barnwood crackle
Paintbrush
Balsa wood
Craft knife
White latex paint
Glue

CRACKLE GLAZE IS A REWARDING PAINT EFFECT.

YOU SIMPLY PAINT A BASE COLOR, APPLY THE GLAZE, THEN ADD YOUR

SECOND CONTRASTING COLOR. THE CRACKS APPEAR IN SECONDS. THE

QUIRKY LITTLE HOUSE ADDS A SENSE OF FUN TO THE RUSTIC BOX STYLE.

1 Paint the candle box cadmium red and let it dry. Paint a base coat of crimson and let it dry fully.

2 Paint a generous, even coat of barnwood crackle on the candle box. Work only in one direction (do not brush back over the crackle glaze). Let the glaze dry for 30–60 minutes.

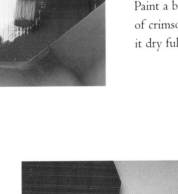

3 Mix one part cobalt blue to two parts aqua paint. Apply the mixture evenly over the dry crackle. Do not brush over the painted area once it has started to crack, as it will not re-crack. Let it dry.

4 Cut a small, simple, house shape out of balsa wood and decorate it. Once it has dried, glue it to the top of the candle box.

Wooden Cache Box

THIS CACHE BOX HAS BEEN DECORATED WITH ONE OF THE QUICKEST PAINT EFFECTS. A CONTRASTING GLAZE IS PAINTED OVER A STRONG BASE COLOR, AND THEN SIMPLY RAG-ROLLED OFF TO REVEAL THE COLOR BELOW.

1 Paint the box white and let it dry. Paint the whole box cadmium yellow and let it dry.

2 Paint one side with the verdigris glaze. Slightly dampen the lint-free cloth and twist it, then begin rolling it across the glaze. Try to roll it in several different directions so that the pattern becomes very smudged. Repeat on all the outside sides, including the base, and let it dry.

3 Paint the top edge of the box with the wine glaze. Cut out the fleur-de-lys stencil on page 125, and with the wine glaze, stencil three fleur-de-lys on each side. Add a little gold stencil powder to each stencil. Let them dry, and then varnish the box.

Plant Container

YOU WILL NEED
Wooden plant container
White latex paint
Translucent glaze in buttercream,
cranberry, and antique gold
Candle
Dragging brush
Steel wool or very fine sandpaper
Varnish

THIS STYLISH, STRIPED, PLANT CONTAINER LOOKS LIKE IT HAS BEEN WORN BY YEARS OF POLISHING. IT IS ACTUALLY A GLAZE DRAGGED OVER CANDLE WAX, WHICH IS THEN RUBBED AWAY TO REVEAL THE BASE COLOR.

1 Paint the whole box white and let it dry. Then paint the whole box with buttercream glaze and let it dry.

2 Rub candle wax on the side edges, top edge, and feet of the box, and on any other area that would generally wear with age.

3 Paint the feet of the box with cranberry glaze and let it dry. Paint large stripes on each side face of the box to the top and bottom edges. Drag them with a brush so that they streak slightly. Do one stripe at a time so the glaze does not dry before you can drag it. When it is dry, paint and drag between the two cranberry stripes with antique gold glaze.

4 Once the whole box is completely dry, gently rub with steel wool or sandpaper so the areas coated in candle wax start to come off, showing the base color underneath. Spray the whole container with varnish.

Apples *and* Pears Wood *Basket*

YOU WILL NEED
Flat wooden basket
White latex paint
Rubber stamps with apple and pear designs
Brush pens
Paper
Scissors
Varnish
Paintbrush

THE DELICATE 3-D DESIGN ON THIS COUNTRY-STYLE, FLAT BASKET IS ACHIEVED BY MASKING AND LAYERING YOUR STAMP WORK. IT CAN BE FILLED WITH FRUIT AND USED TO DECORATE A CORNER IN YOUR KITCHEN OR YOUR DINING-ROOM TABLE.

WOOD

98

1 Paint the basket with white latex paint and let it dry. Then ink the pear stamp with the brush pens.

2 Stamp a pear print on each side of the basket.

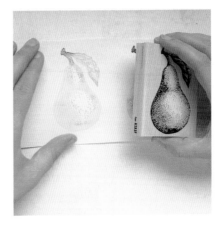

3 Re-ink your stamp when necessary and stamp a pear a couple of times onto paper.

4 Cut out the pear stamps to use as masks.

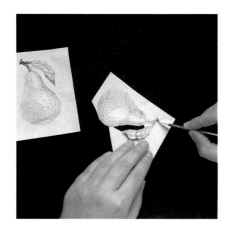

5 Place pear masks over the pears that you have stamped.

6 Ink the apple stamp and print it over the paper used to mask the pears. Stamp apples on a piece of paper, and cut out as before. Then place the apple mask over the stamped apple.

7 Stamp another pear, using lighter tones and colors to add depth.

8 When you have completed the design, remove all the masks and let the ink dry, then varnish the basket.

Marbled Box

YOU WILL NEED
Wooden box with lid
Coral base coat
Terracotta and coral glaze
Thickener
Extender
Plate
Sponge
Veining color
Feather
Spray varnish

Small boxes make charming presents, and this delicately marbled box would make a gift of chocolates extra special, or could be used for jewelry or keepsakes.

1 Paint the box and lid with several layers of coral base-coat paint. Prepare the sponge by wetting it.

2 Squeeze the terracotta glaze into a plate, making random circles and crosses. Repeat with the coral glaze on top of the terracotta. Do the same with the thickener, squeezing three circles, and then again with the extender, squeezing two full circles. Rock the plate slightly to blend the colors and solutions into each other.

3 Dip in the sponge, and start applying paint over the base color on the outside of the box. Do not let the colors get too muddy; if necessary, remake the palette of colors. Leave the inside of the box plain.

4 On a clean plate, squeeze one teaspoon of the salmon veining color. Apply equal amounts of thickener and extender. With a feather, drag the veining mixture over the box, twisting the feather slightly all the time. Soften the veins with a soft paintbrush. Repeat on all sides of the box. Let the paint dry fully, and then spray varnish.

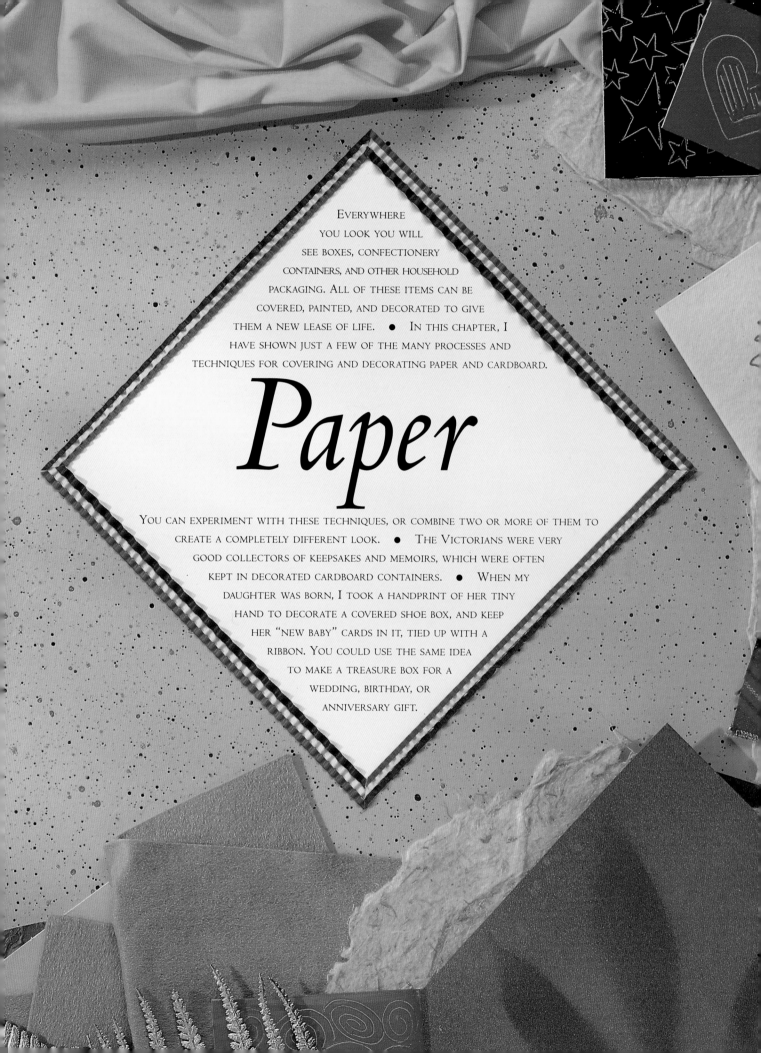

EVERYWHERE
YOU LOOK YOU WILL
SEE BOXES, CONFECTIONERY
CONTAINERS, AND OTHER HOUSEHOLD
PACKAGING. ALL OF THESE ITEMS CAN BE
COVERED, PAINTED, AND DECORATED TO GIVE
THEM A NEW LEASE OF LIFE. ● IN THIS CHAPTER, I
HAVE SHOWN JUST A FEW OF THE MANY PROCESSES AND
TECHNIQUES FOR COVERING AND DECORATING PAPER AND CARDBOARD.

Paper

YOU CAN EXPERIMENT WITH THESE TECHNIQUES, OR COMBINE TWO OR MORE OF THEM TO
CREATE A COMPLETELY DIFFERENT LOOK. ● THE VICTORIANS WERE VERY
GOOD COLLECTORS OF KEEPSAKES AND MEMOIRS, WHICH WERE OFTEN
KEPT IN DECORATED CARDBOARD CONTAINERS. ● WHEN MY
DAUGHTER WAS BORN, I TOOK A HANDPRINT OF HER TINY
HAND TO DECORATE A COVERED SHOE BOX, AND KEEP
HER "NEW BABY" CARDS IN IT, TIED UP WITH A
RIBBON. YOU COULD USE THE SAME IDEA
TO MAKE A TREASURE BOX FOR A
WEDDING, BIRTHDAY, OR
ANNIVERSARY GIFT.

Tartan Box

ADD A TOUCH OF SCOTLAND TO FESTIVE OCCASIONS BY DECORATING A PLAIN BOX WITH CHEERFUL TARTAN PLAID. FILL THE BOX WITH THISTLES AND OTHER DRIED FLOWERS. BLUE AND GREEN HAVE BEEN USED, BUT YOU COULD TRY RED OR PURPLE TARTANS.

YOU WILL NEED
Oval cardboard box
Cream latex paint
Acrylic paint in opaque oxide of chromium, ultramarine blue, and cadmium red
Wide and fine paintbrushes

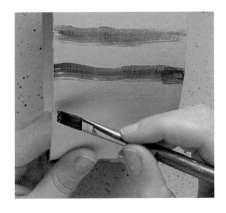

1 Cover the box with several layers of cream latex paint, and let dry. With the opaque oxide of chromium (green), paint vertical lines around the box. Try not to make them completely solid in color, but more dragged in appearance. Then paint horizontal lines around the box at regular intervals. In this case, the box was divided into quarters.

2 Fill the vertical spaces between the green with dragged lines of ultramarine, painting over the green horizontal lines.

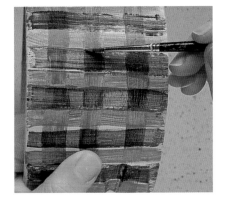

3 Finally, paint a fine, cadmium-red line on each side of the ultramarine.

Chiffon Box

SWAGS OF CHIFFON, LARGE BOWS, AND GOLD TASSELS MAKE THIS A TRULY EXTRAVAGANT BOX. RICHLY COLORED SILK BLOOMS OR GLITTERING JEWELRY COMPLETE THE ROMANTIC DISPLAY.

1 Blend the two violet paints together to make a deep-plum color, then cover the hat box with several coats. Let it dry.

2 Fold the pink chiffon lengthwise to hide the raw edges, then tie it around the hat box. Make a knot and secure it to the box with thumbtacks.

3 Twist the dark purple chiffon to cover raw edges. Secure the end to the back of the box with a thumbtack. Zigzag the dark color up and down the box, tying a chiffon-ribbon bow at each edge, then securing again with a thumbtack. Fix the tassels on the front of the box, hooking them under the thumbtacked bow. Cover the thumbtack points on the inside with removable adhesive.

Sunflower Stamp Box

R UBBER STAMPS WITH VARIOUS DESIGNS ARE BECOMING VERY POPULAR AND ARE VERY SIMPLE TO USE. THIS IS JUST A SHOE BOX, BUT FILLED WITH VIBRANT SUNFLOWERS AND DECORATED WITH THE STAMP DESIGN ON HANDMADE PAPERS, IT WILL BRIGHTEN ANY ROOM.

YOU WILL NEED
Shoe box
Acrylic paint in burnt umber
Stamping ink pens in yellows, oranges, and greens
Sunflower stamp
White watercolor paper
Stamp-ink cleaner
Two green papers

1 Paint the shoe box burnt umber. Paint the sunflower petals with one of the yellow or orange pens. Color the stamens with brown and the mosaic edge in green.

2 Press the stamp firmly onto white watercolor paper. If you have inked the stamp generously, you will make several sunflower impressions. Clean the stamp with the cleaner, then do the same again making petals a different yellow/orange.

3 Tear around the sunflower stamps to form irregular squares. Then tear squares a little larger from the two green papers. Glue a white sunflower square to a plain green square, then stick both to the side of the box. Position three along the long edges and one at each end.

Handmade *Paper Box*

USE SOME OF THE MANY BEAUTIFUL, HANDMADE PAPERS THAT ARE AVAILABLE, AND COVER THE BOX WITH A RANDOM DESIGN. TEARING THE SQUARES INSTEAD OF CUTTING THEM WITH SCISSORS WILL ADD TO THE CASUAL LOOK.

YOU WILL NEED
Round cardboard box
Handmade papers in various colors
Pressed-flower paper
Paper glue

1 Cut out and glue a band of pressed-flower paper to cover the vertical outside edge of your box completely.

2 Tear irregular squares of pale-mauve paper, and glue them around the box at fairly regular intervals.

3 Tear the other papers into ovals, triangles, etc., and glue them randomly around the box. Some of the pressed-flower paper should show through.

THE FERNS AND DEEP-YELLOW OCHER GIVE VICTORIAN STYLE TO THIS BOX. THE FERN TEMPLATE CAN BE OBTAINED FROM A FLORIST. REAL LEAVES CAN BE USED, BUT THEY MUST BE FRESH, NOT DRIED.

Fern Box

114

YOU WILL NEED
Rectangular cardboard box
Acrylic paint in yellow ocher
Paintbrush
Plastic fern
Gold aerosol paint
Dark brown aerosol paint
Aerosol polyurethane varnish

1 Paint the box with yellow-ocher paint. Apply two coats if necessary to cover well, and let it dry.

2 Place the fern along one side of the box, making sure it is flat. Spray gold aerosol paint around the fern and let it dry for a few minutes.

3 Spray dark-brown aerosol paint over the corners, allowing some of the gold to show around the fern. Leave for a minute, then lift off the fern. Spray with polyurethane varnish.

Keepsake *Box*

THIS IS SIMPLY A SHOE BOX COVERED WITH HANDMADE PAPER, TIED AND TAGGED WITH A PRINT OF MY DAUGHTER'S HAND. I MADE IT TO HOLD HER FIRST BIRTHDAY CARDS, BUT YOU COULD ADAPT THIS SIMPLE IDEA FOR A WHOLE HOST OF COLLECTIONS INCLUDING WEDDINGS, ANNIVERSARIES, OR EVEN A SPECIAL HOLIDAY, COVERING THE BOX WITH PHOTOGRAPHS, BUS TICKETS, POSTCARDS ETC.

YOU WILL NEED
Shoe box
Handmade paper
Glue
Bronze acrylic paint
Sheet of plain, white paper
Hole punch
Green garden string

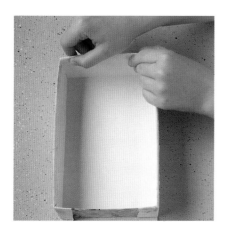

1 Cut out and glue a large piece of handmade paper to the shoe box, covering it completely on the outside.

2 Press several handprints firmly on to plain white paper, using bronze acrylic paint, and allow them to dry. Choose the best hand print for the tag, and roughly tear around it.

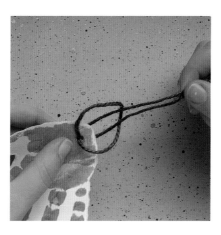

3 Punch a hole into the heel of the handprint, then thread the green garden string through the tag, leaving enough string to be wrapped around the box and tied on top.

Scratch-*card* Box

YOU WILL NEED
Sheet of white cardboard
Hatbox
Scissors
Glue
Three 11 × 16-in. (28 × 43cm)
sheets of metallic-gold posterboard
Acrylic paints in crimson, cadmium
yellow, cadmium orange, cobalt
blue, bright aqua, and white
Matte-varnish spray
Pencil
Paintbrush

USE THIS WILD AND WACKY

BOX TO KEEP SCARVES, NECKLACES,

HAIR ACCESSORIES, AND ALL THOSE

SNIPPETS AND PIECES YOU KNOW ARE

SOMEWHERE! IT HAS BEEN

DECORATED WITH A SCRATCH-CARD

TECHNIQUE.

1 Measure and cut a strip of white
cardboard long enough to cover the
outside of the hatbox. Glue it to the
side of the hatbox and let it dry.

2 Paint blocks of color approximately
6in. (15cm) deep across the width of
one piece of gold posterboard, allowing
the brushstrokes to show slightly. Put
the posterboard aside and let it dry.

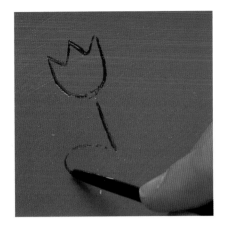

3 On another piece of gold posterboard, paint another strip the same size in crimson, but let it dry only slightly – approximately one minute – until it feels tacky to touch. With a pencil, start to scratch out a design – for example, a tulip – so that the gold board shows through the paint. Repeat this sketch, leaving 1in. (25mm) between each flower. Let it dry. Repeat this technique with all the acrylic colors, scratching different designs, such as hearts and stars, onto each color.

4 Cut out some of the designs, leaving a small border around them. Cut other designs into rectangles and squares with the design in the center – this has been done with the daisy and the tulips. Cut the patches of plain color so they can be used as backgrounds to set off the cut-out designs, such as an emerald-green square to contrast with the red tulip design, or a cadmium-yellow rectangle for the orange watering can.

5 Glue the scratch-card designs onto the contrasting squares and rectangles of color, and then let them dry.

6 Begin gluing the different shapes and designs onto the white cardboard that you glued around the hatbox. Try to create a haphazard design by overlapping some pieces and mixing the contrasting colors.

7 Continue building up the design, trying to fill as much of the white cardboard as possible with colorful shapes, until you have completely covered the outside of the box. Then, let it dry.

8 Finally, spray the decorated box with aerosol-matte varnish for protection, and let it dry.

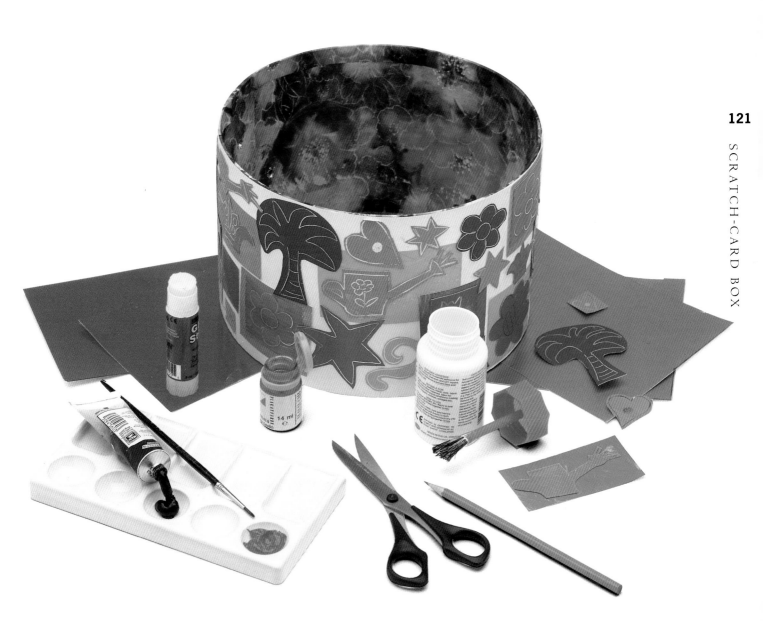

Templates

ALL OF THE TEMPLATES SHOWN ON THE FOLLOWING PAGES ARE ENLARGED BY 200 PERCENT.

THESE ARE JUST SUGGESTIONS; YOU CAN, OF COURSE, CREATE YOUR OWN.

TO REDUCE THE DESIGNS, USE A PHOTOCOPIER SET AT 50 PERCENT. ALTERNATIVELY, YOU

CAN DRAW A GRID OF SQUARES ON THE DESIGN AND THEN COPY EACH SQUARE ONTO A GRID

OF SMALLER SQUARES. FOR EXAMPLE, 1-INCH SQUARES COULD BE COPIED ONTO 1/2-INCH

SQUARES TO MAKE THE IMAGES HALF THE SIZE SHOWN HERE.

Butterfly motif
Stenciled Flowerpot
(p. 18)

Ivy motif
Stenciled Flowerpot
(p. 18)

Daisy motif
Stenciled Bucket
(p.38)

Iris motif
Stenciled Bucket
(p.38)

Tulip motif
Stenciled Bucket
(p.38)

Heart motif
*Hearts and Flowers
Basket*
(p.80)

Daffodil motif
*Hearts and Flowers
Basket*
(p.80)

Fleur-de-Lys
motif
Wooden Cache Box
(p.94)

Star motif
Stars and Stripes
(p.88)

Index